YOUR KNOWLEDGE HAS VALUE

Noel Mwenda

Is development of Dodoma Municipality sustainable? How sustainability is?

GRIN Publishing

Bibliographic information published by the German National Library:

The German National Library lists this publication in the National Bibliography; detailed bibliographic data are available on the Internet at http://dnb.dnb.de .

Imprint:

Copyright © 2011 GRIN Verlag, Open Publishing GmbH
Print and binding: Books on Demand GmbH, Norderstedt Germany
ISBN: 978-3-656-17322-9

This book at GRIN:

http://www.grin.com/en/e-book/191469/is-development-of-dodoma-municipality-sustainable-how-sustainability-is

GRIN - Your knowledge has value

Since its foundation in 1998, GRIN has specialized in publishing academic texts by students, college teachers and other academics as e-book and printed book. The website www.grin.com is an ideal platform for presenting term papers, final papers, scientific essays, dissertations and specialist books.

Visit us on the internet:

http://www.grin.com/

http://www.facebook.com/grincom

http://www.twitter.com/grin_com

Is development of Dodoma Municipality sustainable? How sustainability is?

By

First Author:

Canicius J. Kayombo

Seneour Botanist

Institute of Rural Development Planning

&

Second Author:

Noel Mwenda

Assistant Lecturer

Department of Psychology

School of Educational Studies

The University of Dodoma

Is development of Dodoma Municipality sustainable? How sustainability is?

ABSTRACT

The study was conducted at the area from Institute of Rural Development Planning and Management to Dodoma Municipality centre to assess whether development can be sustainable and how this sustainability is maintained. An organized walk was set along the road and plot was designed at an interval of 200 meters. Any observed condition leading to unsustainability was recorded, settlement type, vegetation types, drainage systems at IRDP and roads, livestock and crops found were recorded. Computer programs including macro-soft excel, macro-soft word were used during data collection and analysis. The study revealed that even the drainage systems and roads around higher institutions like IRDP were not well maintained. The road maintenance including concrete and tarmac road had some implications of not encouraging sustainable development as drainage systems were not directed and maintained properly, off roads were prominent at the earth road from IRDP to Mwisho wa Lami around Miyuji area. The area was also, occupied by mostly exotic trees and indication that the natural vegetation scenery is no longer existing. Overgrazing around settlement and open spaces found along the whole road distance of 5km was very alarming done mostly during weekends, after office working hours and night hours, an indication that land use suitability is not implemented and hence not sustainable development. Sewage systems are not constructed around Miyuji area and thus the community is vulnerable to deadly diseases like diarrhea and Typhoid.

The municipality is expanding rapidly through buildings, industrial construction and more traffic emitting deadly gases also, contributing to global warming. This research recommends that more survey should be conducted to reveal more human actions that lead to an alarming negative effects resulted by development, indigenous trees should be planted to maintain natural vegetation scenery, drainage systems at homesteads should be constructed and this can be done involving the local people and government through sharing costs and again must be implemented as soon as possible, institutions which are the areas believed to accommodate knowledgeable people should be a model to the

2

community, grazing at the municipality centre should be discouraged to rescue the ornamental plants that also, play a role as carbon gases sink and maintain the municipality beauty, regulate temperature, encourage reliable rainfall, prevent erosion and provide oxygen of which without, human survival is impossible.

KEYTERMS: Development, Sustainable, Surrounding community

INTRODUCTION

Background information

Sustainable development

This is a social, political and economic growth of the present generation in the way that does not compromise the needs of future generations. Sustainable development should value the none living component, scenery as well as the richness of biological diversity. Members of living organisms have strong effect on each other (Castro & Huber, 2005), and none of them can survive by its own. Sustainable development embraces everything surrounding human beings (Dupriez & Leener, 2003).The living environment includes plants and animals which their existence are mostly dependent on the human's generosity (Dupriez & Leener, 2003).

Worldwide, the human population is competing for resources as well as improved life through advancement of settlement, roads, industries and food supply. Environmental crisis has been experienced all over the world since the second world war when several damage on environment was experienced. After the war, the only left struggle was for resource accumulation (ReVelle & ReVelle, 1988), to fill the social and economic satisfaction needs.

Tanzania is one of African countries with a large number of people facing severe poverty rate and thus depending to a large scale on any abiotic and biotic resources available around them. Also, the fast emerging implementation of infrastructure like roads has been found to bring about more negative effect than the positive part of it. Development lead to pollution of water, deforestation as a result of land use dynamics and human pressures; most people depend on woody materials for fuel (Assefa, *et al*, 2001).

Development always alters the biodiversity habitats and some plants and animals have become endangered (ReVelle & ReVelle, 1988). Under this situation, sustainable development cannot be achieved unless the human generation becomes ready to implement the sustainable development plans.

Brief historical change of Dodoma from the previous up to now

Dodoma municipality has developed gradually from a very small size to how it is now. Historically, the area was covered by natural shrubs with scattered trees of a variety of species including *Acacia nilotica, Acacia tortilis, Acacia polyacantha, Ficus sur, Ficus sycomorus, Markhamia obtusifolia, Erthrina abyssinica*. This vegetation accommodated a diversity of wildlife such as dik dik, lions and birds. The expansion of settlement has called for initiation of various government offices, establishment of small industries, road transport improvement. At the same time any wildlife trying to go across is killed. Soil erosion is prominent and thus soil degradation. The existing vegetation is completely new, comprising exotic trees such as *Eucalyptus* sp., *Cryostegia madagascariensis, Jacaranda mimosifolia, Senna siamea, Senna spectabilis, Nerium oleander, Thevetia peruviana* and many others not mentioned in this experimental study. Beentje (1994) pinpointed that the natural vegetation is severely degraded due to settlement establishment, and when a need for vegetation cover occurs, it is complemented with non-native trees. This is not sustainable development.

Development Sustainability Possibilities and how sustainable we are

Apart from the human being's struggle for sustainable development, this idea cannot be fulfilled easily. Arms (1996) outlined that the world cannot continue to feed the world's population without continuing to deplete the world's resources and that low input farming without using a lot of energy pesticides, fertilizer and water will sustain environment but will mean less production or crop output. This implies that modern agriculture to human beings means application of a lot of chemicals that will enable our food plants grow faster and produce more than we expected previously no achieve the best goal.

Economic growth creates its own ruin to environment and sustainable development has been criticized as an ambiguous idea with a wide range of interruption, many of which are contradictory (2002). There is no way that we can sustain environment while there is an increasing need for improvement infrastructure and population growth with enormous needs for renewable and non-renewable resources.

Challenges facing Sustainable Development
Among many others, there is population growth; this needs to be fed by agriculture resources produce from large areas, specifically land. Social structure is also a major obstacle to sustainable development in many countries as it gives most of the nation's wealth to a tiny monitoring of its people (Smith, 2002).

Advancement in industries and transport systems; the big industries require large vehicles that will be involved in carrying heavy luggage to various areas and hence more roads construction lead to damage of land biodiversity (Smith, 2002).

Resources ownership; smith (2002) argued that when every body shares ownership of a resources there is a strong tendency to exploit and misuse the resources, of which sustainability becomes a bizarre.

METHODOLOGY

Study area

The study was conducted along Dodoma to Kondoa road; from Institute of Rural and Development Planning to Down town Dodoma Municipality.

Land use type

The land use include settlement, small scale farming around homesteads, grazing, various institutions such primary schools, constructed higher learning institutions such as Institute of Rural Development Planning, Capita teachers training college, various religious institutions, communication through road, and small open spaces..

Vegetation type

Woodlots,

These are clusters of trees established by households at homesteads as source of firewood, shade, windbreak and ornamental function.

Thicket-woodlands

Rare natural vegetation patches within the open spaces that mainly accommodate indigenous plants, to mention few are; *Dicrostachys cinerea, Acacia tortilis, Commiphora africana, Vangueria infausta, Turraea robusta, Ximenia Americana, Maerua* spp and *Euphorbia tirucali* which is also used as traditional fence and boundary mark.

Crops

The grown crops include *Zea mays* (Maize), *Phaseolus vulgaris* (Beans), *Saccharum officinarum* (Sugarcane), *Hibiscus esculentus* (Bamia), *Lycosperscon esculenta* (Tomatoe), *Cajanus cajan* (Pigeon peas), *Ipomoea batatus* (Sweet potatoe), *Alium cepa*

6

(Onions), *Vitis vinifera* (Grape), *Arachis hypogaea* (Ground nut), *Pennisetum glaucum* (Millet).

Livestock keeping

Animal husbandry includes cows, goats, sheep, pigs, hens, ducks and guinea fowls.

Methods and materials

A walk was organized along main road to Kondoa from Dodoma Municipality centre at a distance of about 5km, whereby the IRDP constructed are was observed including the roads based on conditions either tarmac or not. The observed conditions were recorded such as off roads, drainage systems. The found vegetation types were described and crops food identified. Any found animals were recorded. Stationeries were employed during data collection. Botanical books were used to justify the scientific names, also, other literatures for correlation of the observed situation and other researcher's views. Computer programs, especially macro-soft excel, macro-soft word and accessories were used during compilation and analysis of data.

Data analysis

The collected data was analyzed through macro-soft and macro-soft excel.

RESULTS AND DISCUSSION

Construction at various institutions such as at Institute of Rural Development Planning and Management (IRDP)

The construction at IRDP, just like in other areas of the world does not ensure sustainability in many ways. Rain water and sewage drainage systems are working efficiently and thus bringing about high water accumulation around the campus. This can lead to eruption of malaria due to available breeding sites for mosquitoes.

7

This can be concluded that good road encourages heavy traffic that at the end leads to emission of waste gases that finally lead to global warming.

Deterioration of biodiversity and natural vegetation has been noted to be a prominent effect and hence sustainability is impossible.

Road construction from IRDP to Mwisho wa Lami

Its construction is not that much based on sustainability, it is very rough in such a way that off road traffic is very common and thus affecting the vegetation by trampling and or plant diversity is degraded. Corrugations and ditches of water are very common; erosion and or land degradation is common. Soil exhaustion refers to as being used or collapses completely (Collin, 2001) and thus become unsuitable for a particular purpose.

Mwisho wa Lami to Dodoma municipality centre

Along road plant diversity has been destroyed due to settlements and several off roads. Poorly maintained road side ridge lead to erosion. Rain water drainage system poorly constructed and thus accumulation of water affects the settlement areas. Most businesses are pooled to the roadside and thus leading to pollution resulted by throwing wastes.

Development of Dodoma town construction industry relating to sustainable development

Historically, the area was covered by natural vegetation that could have been cleared to give a room for construction and thus the fauna composition was damaged as well as land degradation the scenery has completely changed. Congestion and quality are growing problems. Traffic is chaotic almost all the time and can lead to expansion of roads (Cunningham *et al*, 2007). Urbanization is very prominent in rapidly growing areas where there is heavy traffic

Sustainable development is evidently seen as an impossible completely fulfilled matter because of the environmental pollution (Cunningham *et al*, 2007)

8

CONCLUSION AND RECOMMENDATIONS

The development is thought to be sustainable theoretically even though in practical sense seem to be un attainable miracles. Any activity done by human being mostly has positive effect when satisfaction of needs is targeted, although in long run the turns to bring negative effects. The vegetation being cleared as well as degraded soil will lead to severe challenges and measures will not be in the position to overcome the problems as easy as possible.

Proposed measures to sustain environment at IRDP to Downtown Dodoma Municipality

The measures that could help to sustain environment include;

1) plant more trees, control soil erosion, construction of damaged system that will allow rain water to flow out efficiently as for now when it rains, most water is accumulated between library and administration block. This may lead to eruption of diseases such as malaria and cholera.

2) The road to Dodoma town should be maintained in the way that erosion is protected.

3) This study sets recommendations that construction should respond to recommended standards. Roads should be constructed well to avoid off road traffic.

4) The natural vegetation should be set aside and protected for sustainable biological and non-biological component.

5) Land use should be compatible such as human settlement should mean that and that for cattle keeping should be want it means, where as crop farmers must be separated from cattle keepers.

6) Reasonable distance from road and settlement must be taken into account to improve road side environmental sustainability.

REFERENCES

Arms, K. 1996. Sustainable agriculture. Holt environmental science United States of America.

Asefa, T., Rugumamu, S.M. & Ahmed, A.G.M. 2001. Globalization, democracy, and development in Africa challenges and prospects. Organization for science research in Eastern and South Africa.

Beentje, H. 1994. Kenya trees, shrubs and lianas. National Museum of Kenya, Nairobi-Kenya. Worldwide fund, USA.

Castro, P. & Huber, M.E. 2005. Biodiversity. All creatures great and small. Marine biology. The McGraw. Hill higher education. United States of America.

Collin, P. Dictionary of Ecology and Environment. Fourth edition. Peter publishers, Britain.

Cunningham, W.P. Cunningham, M.A. & Saigo, B.W. 2007. Environmental science. MacGraw hill companies. MacGraw Hill high education-United States of America.

Dupriez, H. & Leener, P.D. 2003. Agriculture in African rural communities. Crops and Soils. Belgium.

ReVelle, P. & ReVelle, C. 1988. The environment issues and choices for society. Third Edition. Printed in the United States of America.

Smith, E. 2002. Economic growth and sustainable development. Environmental science. A study of interrelationships. McGraw hill higher education.